The Accelerated
Parent

The Accelerated Parent

How ADHD Can Overtake Evolution

Knut Knott

The Accelerated Parent
How ADHD Can Overtake Evolution

ISBN 979-8-328-74166-8

To my wife.

Table of Contents

Introduction

Attention Deficit Hyperactivity Disorder (ADHD) is a highly heritable condition characterized by impaired dopamine regulation in the brain. Dopamine plays a crucial role in regulating mood, attention, and self-control. When dopamine levels are insufficient, these functions can be disrupted, causing individuals with ADHD to struggle with initiating or completing tasks. They may also experience a lack of impulse control and frequent urges for physical movement, even when unnecessary or inappropriate.

Due to its well-understood nature, ADHD is one of the most medically treatable disorders. Stimulant medications can increase dopamine levels in the brain, helping individuals regain much of the function impaired by the condition. This treatment can dramatically improve focus and attentiveness, and the physical restlessness may disappear entirely while the medication is active.

Given the heritable and medically treatable nature of ADHD, an intriguing question arises: why hasn't evolution eliminated this condition? Typically, traits that increase the likelihood of premature death are gradually eliminated through natural selection. There is a strong case for ADHD being a survival disadvantage: for children, impulse control is vital to avoid fatal accidents; young adults need impulse control to maintain social boundaries and avoid violent confrontations; and for adults, executive skills are crucial for planning and managing the resources necessary for survival. However, the persistence of ADHD suggests it may have unexpected evolutionary advantages.

One popular theory suggests that traits like heightened impulsivity and novelty-seeking behavior could have been beneficial for hunters or nomads in a prehistoric context. Yet, this model faces significant challenges, such as the incompatibility of the restless body movements common in ADHD with the stealth required for successful hunting.

This mini-book argues that the traditional understanding of ADHD has been limited by an evolutionary perspective focused mainly on survival, overlooking the complexity of human evolution. For human males, evolutionary fitness is predominantly about traits that enhance their ability to attract mates and seize opportunities to procreate. However, these traits associated with sexual selection and reproduction do not necessarily align with those needed for survival.

Human females have an even more complex role in evolution. They typically do not face a shortage of potential mates and have the flexibility to select less optimal partners if necessary, allowing them to control their reproductive rate and strategy. Understanding the evolutionary fitness of ADHD requires recognizing the distinct ways in which males and females exhibit the condition and their complex roles in evolution.

By highlighting a unique reproductive strategy inherent to individuals with ADHD, it will be demonstrated how these traits can confer a seemingly confounding yet surprisingly effective reproductive advantage over the general population.

Hypothesis Overview

The main hypothesis of this mini-book is that ADHD enhances evolutionary fitness by shortening the generational interval. Research indicates that individuals with ADHD are significantly more likely to become young parents. Although this research was conducted in a modern context, it is plausible that prehistoric humans with ADHD exhibited similar tendencies in their mating behaviors. There is no evidence to suggest that the underlying dynamics of ADHD-related reproductive strategies have changed significantly over time.

In a growing population, reducing the time between generations leads to faster growth. When parents have children at a younger age, the population size increases more rapidly. This happens even if each family has the same number of children. The key factor is the frequency of generations: more frequent generations mean more people in a shorter period.

While this might seem counterintuitive—since we often think population growth depends solely on the number of children per family— the timing of births is equally important. Younger parents lead to more generations within the same timeframe, boosting the growth rate over time. A detailed explanation of this mechanism will be provided in the third chapter.

From this perspective, ADHD enhances evolutionary fitness not by being inherently beneficial to the carrier of the trait, but by acting as a brute force mechanism through which nature sacrifices mental well-being to accelerate reproduction. By impairing men's executive function skills, they are predisposed to seek adventure and engage in risk-taking behaviors, which often include an early interest in sexual activity. Men with ADHD struggle to pursue or follow through on personal goals and dreams that might delay reproduction.

Similarly, by impairing a woman's executive function skills, ADHD hinders her ability to pursue or follow through on personal goals and dreams that might delay reproduction. This impairment also reduces her capacity to track menstrual cycles, control impulses related to family planning, or engage in a prolonged search for the ideal partner, thereby minimizing the potential loss of reproductive years.

While this harsh twist of nature may seem grim or unsatisfactory, it offers a plausible explanation for why males often experience more severe impairments than females. In nature, females with ADHD need to remain highly functional to ensure the survival of their offspring. In contrast, males with ADHD may exhibit more severe impairments because risk-taking is a primary reproductive strategy. The additional selective pressure on mothers with ADHD means that young females must maintain a higher level of functionality for nurturing roles, while males with ADHD can afford to be more impulsive and reckless since their primary concern is spreading their genes, relying on females to care for the offspring.

For both sexes, ADHD presents significant challenges to mental well-being. However, evolution is neither fair nor kind; it simply favors those who produce the most offspring in the shortest possible time. If ADHD accelerates the generational cycle, it can enhance reproductive fitness, even if it results in a lifetime of chaos, anxiety, and troubled relationships. Evolution prioritizes the perpetuation of genes over the quality of life of the individuals carrying them.

ADHD and Young Pregnancy

In 2019, a Swedish research group from Uppsala University and Karolinska Institutet published a national cohort study examining the link between teenage births and ADHD in girls. The study analyzed data from over 380,000 Swedish women who gave birth between 2007 and 2014 and concluded that girls with ADHD were six times more likely to become teenage mothers compared to their non-ADHD peers (JAMA Netw Open[1]).

In 2006, Flory et al. conducted a study exploring the correlation between childhood ADHD and risky sexual behaviors in young men aged 18 to 26. The study found that young men with ADHD were six times more likely to have impregnated someone by that age, indicating that the predisposition for early reproduction is not limited to females (JCCAP[2]).

---

1 Association of Attention-Deficit/Hyperactivity Disorder With Teenage Birth Among Women and Girls in Sweden. Retrieved from https://www.ncbi.nlm.nih.gov/pmc/articles/PMC6777395/

2 Childhood ADHD Predicts Risky Sexual Behavior in Young Adulthood. Retrieved from https://www.tandfonline.com/doi/abs/10.1207/s15374424jccp3504_8

The connection between ADHD and teenage pregnancies is well-documented. Individuals with ADHD are more susceptible to impulsive decision-making and risky behaviors, such as early sexual activity and unprotected sex. This impulsivity, coupled with challenges in planning and using contraceptives consistently due to inattention, significantly increases the risk of unintended pregnancies. Furthermore, girls with ADHD are more likely to carry their pregnancies to term rather than opting for abortion.

Since ADHD is a chronic condition, there is no evidence to suggest that the tendency for early reproduction disappears upon transitioning into adulthood. Indeed, research reveals a strong link between maternal age at first birth (MAFB) and the likelihood of ADHD in offspring. While one might hypothesize that the mother's age influences the risk of ADHD in children, studies suggest that this relationship is largely explained by genetic factors. Specifically, individuals with ADHD are more likely to have children at a younger age, passing on the genetic predisposition for ADHD to their children. Thus, the early maternal age associated with ADHD in children primarily reflects the young parent's ADHD, which they transmit to their offspring (Oxford Academic[3]).

Although these studies were conducted in modern times, it is plausible that prehistoric humans faced similar family-planning challenges. For women, childbirth has always been a significant long-term commitment with considerable risks,

---

3 Maternal age at childbirth and risk for ADHD in offspring: a population-based cohort study. Retrieved from https://academic.oup.com/ije/article/43/6/1815/709366

including the risk of death. Family planning and contraceptive strategies have likely been crucial aspects of women's lives since prehistoric times. It is conceivable that ADHD may have similarly impaired these strategies for prehistoric women, making it more challenging for them to effectively plan and space their pregnancies.

Similarly, the male traits that lead to early pregnancies are likely timeless. There is no evidence to suggest that risk-taking behaviors and an early interest in sexual activity are confined to modern culture. These traits are probably rooted in fundamental aspects of human behavior and biology that have persisted across different eras and societies.

Population Growth and Generational Intervals

When determining the evolutionary fitness for a species or genetic trait, traditional evolutionary science will often emphasize the number of offspring and their survival skills. These are seen as the primary factors determining the evolutionary success of those genes. It is typically assumed that the generational interval is constant, which is the average time between the birth of parents and the birth of their offspring. The generational interval is an often-overlooked feature that is, in fact, equally important as the number of offspring. If the generational interval varies between individuals within a species, it must be taken into account.

This book hypothesizes that the primary feature of ADHD genes is to manipulate the generational interval, exerting pressure on individuals carrying those genes to become young parents. This chapter will demonstrate the mathematical effects of changing the generational interval. For the sake of argument, let's assume that the general population has an average generational interval of 25 years, while individuals with ADHD have a shorter average generational interval of only 20 years. These figures are hypothetical but align with scientific literature, which indicates that the maternal age at first birth can vary by several years between women with ADHD and the general population. We will also assume that ADHD is perfectly hereditary and that the population doubles with each generation, corresponding to a family of four children (who all survive).

Figure 1 illustrates the generational tree of the ADHD lineage compared to the general population over a span of one hundred years. The ADHD lineage experiences five generational shifts, while the general population undergoes only four. Consequently, at the one-hundred-year mark, the ADHD lineage's current generation consists of as many as 32 individuals, whereas the general population is one generation behind, producing only 16 individuals.

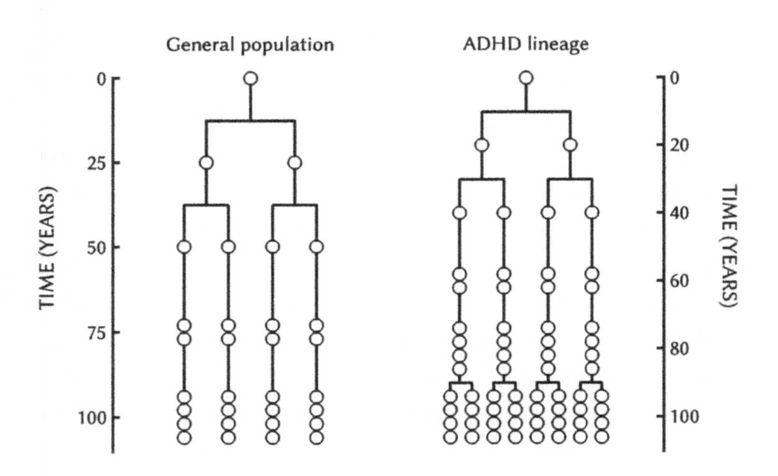

Figure 1: Generational tree of ADHD lineage

Let's also illustrate this population expansion with a plot. Figure 2 compares the population expansion of the ADHD lineage to that of the general population. The circles and squares denote the generational shifts, which are more closely spaced for the ADHD lineage. The number of offspring per family is not higher in the ADHD lineage; it is only the timing of the generational shifts that is more closely spaced. Still, the *annual* growth rate is much greater for the ADHD lineage, since more generations pass over the same timeframe.

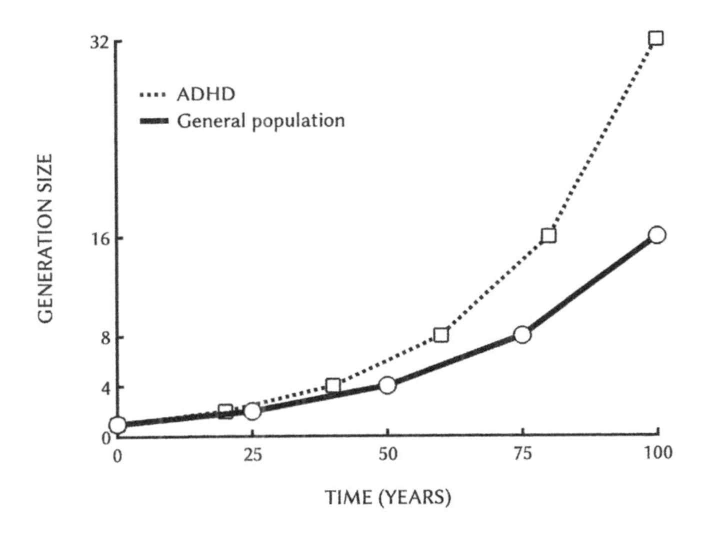

Figure 2: Population growth of ADHD lineage

The figure clearly shows how both lineages grow exponentially, although the ADHD lineage grows at a much faster rate. The growth rate per generation is the same, but the *annual* growth rate differs. In this particular example, we can easily calculate the annual growth rate from the known doubling time, which corresponds to the time for one generation to pass. The calculated annual growth rates are as follows:

$$r_{\text{ADHD}} = 2^{\frac{1}{20}} \approx 1.035 \ (3.5\%)$$

$$r_{\text{GenPop}} = 2^{\frac{1}{25}} \approx 1.028 \ (2.8\%)$$

The difference in *annual* growth rates is significant: 3.5% for the ADHD lineage compared to 2.8% for the general population. The growth advantage of having your offspring five years earlier is almost as great as having one more child per female. Clearly, the generational interval should be taken seriously when measuring evolutionary fitness. Even if the effect of ADHD on the generational interval is much less than five years, perhaps only a few months, the population expansion over thousands of years would be significant—more than enough to counteract some survival disadvantages.

For clarity, our example asserts no selective pressure against the general population. The general population produces an identical number of offspring per family and has survival skills as competent as the ADHD lineage. Yet, the general population is simply outnumbered by brute force by a genetic lineage that suffers greater survival challenges, but which broke the code of natural selection by reproducing faster, outpacing everyone to win. This paradox underscores the substantial impact of the generational interval on population growth and illustrates the complexity of evolutionary mechanics.

The ADHD-Dominant Society

The explosive population growth of the ADHD lineage suggests that ADHD could potentially overtake the human species and become the dominant form. However, if everyone carried these traits, the societal effects of ADHD might be greatly magnified. ADHD males are often predisposed to disorder, impulsivity, and sometimes violence. It is possible that a society dominated by individuals with ADHD could be fundamentally dysfunctional. Without a functional hierarchy where people work together toward common goals and respect boundaries, such a society might resort to violence and internal conflicts over scarce resources and mating opportunities. Furthermore, individuals with ADHD often have a strong aversion to monotonous and repetitive tasks, which could lead to significant male competition to avoid more labor-intensive roles within the social hierarchy. These behavioral challenges might ultimately limit the sustainability of a society dominated by people with ADHD. It seems likely that a balance with the general population would be necessary to take on the boring tasks and essential roles in resource management and executive function.

Moreover, in life-or-death competition against other humans, such as in conflicts or warfare, executive functions are likely crucial for strategic success. Anticipating the moves of your enemies becomes easier if they lack the discipline, impulse control, or patience needed to follow through on strategies and endure a drawn-out standoff. It is conceivable that having a variety of personalities within your group, including both impulsive individuals and cautious strategic thinkers, can provide a significant advantage in wartime. This mix of personalities allows for a passive-aggressive strategy that is both fierce and difficult for opponents to anticipate.

A similar argument applies at the family level. Partners from the general population may be crucial for individuals with ADHD to lead a successful life. A mother with ADHD might significantly boost her children's survival chances by relying on a partner from the general population to handle the routine, daily tasks of gathering resources and providing structure. Similarly, an ADHD father might need a highly functional partner to provide emotional stability and care for their children. In couples where both partners live with untreated ADHD, the survival disadvantage may simply be too great. If that holds true, selection pressure would maintain a low enough prevalence to make dual-ADHD pairs sufficiently unlikely to form. This idea is speculative at best, but it supports the observed prevalence patterns.

This chapter might give the impression that ADHD features are useless, but that is far from the truth. On the contrary, ADHD encompasses a spectrum of personal traits that include highly desirable qualities such as daring, fun, adventure, creativity, and spontaneity. While these traits are not always ideal for survival, they can make life more enjoyable and open doors to opportunities that might be missed by always playing it safe and following the lead of the "boring" spouse or colleague. This dynamic underscores the importance of complementary roles to ensure both stability and progress, highlighting the evolutionary significance of diverse traits within a population.

Those who suffer the most from ADHD are often the individuals who have ADHD themselves, as it frequently coexists with conditions such as depression, obesity, insomnia, educational failure, and a shortened lifespan. However, those around a friend with ADHD can enjoy their spontaneity and adventurous spirit when it suits them, set boundaries against the riskiest activities, and retreat to their own orderly lives at the end of the day. The person with ADHD does not have that luxury but lives with the condition non-stop, possibly requiring stimulant medications to establish healthy boundaries and manage themselves in private settings.

Nature might be so crude that adventure-seeking traits were meant only to enhance reproductive speed. However, this doesn't contradict the idea that traits associated with ADHD have played a crucial and positive role in human development and the qualities that set us apart from other animals. When a segment of the population continually challenges the status quo, it can lead to innovation and the exploration of new ideas and territories. This drive for novelty and risk-taking could push humanity to make discoveries and advancements that might otherwise be unattainable. If everyone possessed perfect self-control and risk aversion, we might never have started to eat meat, use tools, create art, or venture beyond our early habitats. Thus, ADHD traits such as impulsivity and curiosity could have been essential in driving human progress and exploring new frontiers.

Hunters and Farmers

Proposed by Thom Hartmann in the 1990s in his book "Attention Deficit Disorder: A Different Perception," the Hunter vs. Farmer Hypothesis offers an intriguing explanation for the persistence of ADHD traits in the human population. This hypothesis suggests that the characteristics associated with ADHD were once advantageous in certain prehistoric environments. According to the hypothesis, human societies originally consisted of two primary survival strategies: hunting and farming.

Hunters needed to be impulsive, quick-thinking, and capable of hyper-focusing on their immediate surroundings. These traits allowed them to respond swiftly to threats, track prey over long distances, and react to sudden changes in their environment. Conversely, farmers required patience, long-term planning skills, and the ability to engage in repetitive tasks, such as planting and harvesting crops. The hypothesis proposes that the traits associated with ADHD could have been beneficial for hunters. Hyperactivity and impulsivity could translate into quick responses to dangers and opportunities, and the ability to hyperfocus would be beneficial when stalking prey or navigating complex landscapes. Moreover, novelty-seeking behaviors would drive exploration and innovation, crucial for finding new food sources and adapting to changing environments.

While intriguing, this hypothesis raises several questions. Many who have struggled with their own ADHD or cared for a family member recognize it as a significant impairment. It's reasonable to wonder if ADHD could be beneficial for activities requiring sustained attention, such as foraging for food or hunting prey every day. There is also a conflict between the hyperactivity aspect of ADHD and the stealth and patience required for successful hunting. Hunting and fishing often involve considerable waiting and stillness. For example, to catch a fish, you don't jump into the lake and swim around; instead, you find a strategic position and remain quiet until the fish comes to you. Similarly, foraging through a forest without causing noise from branches and foliage is difficult. Hunting usually involves finding a strategic location where prey may pass by if you wait long enough. This level of patience and stillness is something individuals with ADHD might find particularly challenging to maintain.

Despite these challenges, it's worth noting that the Hunters vs. Farmers hypothesis doesn't claim ADHD traits are universally advantageous. Rather, it suggests that certain aspects of ADHD, such as heightened alertness and quick reaction times, could have been beneficial in specific contexts. This nuanced view recognizes that while ADHD traits might have offered benefits in some situations, they could be detrimental in others. What might be called into question is the evolutionary fitness of ADHD traits—the idea that they are beneficial enough to confer a net survival gain despite significant impairments in other areas of life. It is reasonable to consider if social and executive functions may be greater survival skills than heightened awareness. For ADHD traits to contribute to evolutionary fitness, their survival advantages must outweigh the disadvantages. Understanding this balance is crucial for determining whether the traits associated with ADHD could have been selected for in human evolution.

Attempts have been made to trace ADHD-associated genes through ancient DNA samples to test the Hunters vs. Farmers hypothesis. Farming began around 10,000 to 12,000 years ago, but ancient human DNA samples go back as far as 45,000 years, well into the hunter-gatherer era, making it possible to study genetic shifts in that period. A study published in Scientific Reports by Reiersen et al. analyzed these ancient samples and concluded that ADHD-associated genes have been steadily declining for at least the past 45,000 years, long before humans transitioned to settled agricultural lifestyles. This decline suggests that natural selection acted against ADHD genes even before humans had invented farming, contradicting the idea that these traits were advantageous for the survival of hunter-gatherers (Scientific Reports[4]).

While the study showed a small but steady decline of ADHD-associated genes in ancient samples, it does not contradict the theory that ADHD traits may have once been beneficial but later became maladaptive due to environmental changes, provided those changes occurred more than

---

4 Genomic analysis of the natural history of attention-deficit/hyperactivity disorder using Neanderthal and ancient Homo sapiens samples. Retrieved from https://www.nature.com/articles/s41598-020-65322-4

45,000 years ago. However, since current ancient DNA samples date back no more than 45,000 years, this hypothesis cannot be confirmed or refuted at this time.

Now, consider the hypothesis proposed in this book: ADHD does not provide a survival advantage but merely shortens the generational interval. This would mean that the prevalence of ADHD features remains relatively constant over time, representing a balance point. At this balance point, the benefit of faster reproduction associated with ADHD is counteracted by the survival disadvantages it imposes on those who carry it. Consequently, natural selection neither eliminates ADHD genes nor allows them to dominate the general population. Nonetheless, minor fluctuations in prevalence might occur over time.

In the early days of human population expansion, when resources were abundant and nature was our primary threat, the faster reproduction rate conferred by ADHD could have been a significant advantage, facilitating the spread of those genes. However, as human societies evolved and competition for resources with other groups intensified, executive function likely became more crucial for survival, given the increased importance of strategic thinking and planning in these intergroup conflicts. ADHD features that exerted pressure to reproduce faster during the early human population expansion may now be negatively selected in the era of inter-human competition.

It must be emphasized that this is only a theory proposed by this book. However, it is consistent with ancient human DNA samples. There have been minimal changes in the prevalence of ADHD genes over the past 45,000 years, and if anything, these genes appear to have been a minor survival disadvantage, at least since human populations became abundant and developed tool-making and complex social structures approximately 50,000 years ago. It is possible that the presence of ADHD traits in ancient DNA samples from 45,000 years ago might be better explained by their rapid spread during early population expansion as humans conquered the world, followed by a steady decline due to increased competition among humans.

Group Fitness vs. Individual Fitness

When evaluating the evolutionary fitness of a trait, it's important to discern whether it primarily benefits the group as a whole or specifically promotes the genes of an individual lineage within the group. This distinction can affect how the trait presents itself and the mechanism by which it is inherited.

For example, a trait like adventure-seeking might benefit the tribe by fostering exploration and innovation. However, it also puts the individual at significant risk of premature death. Logically, a hereditary predisposition for adventure-seeking would quickly be eliminated by natural selection because individuals exhibiting this trait would have a lower survival chance than the general population. This paradox highlights the complex interplay between individual risk and group benefit in the evolution of certain traits.

To illustrate group fitness, we will examine one of the most universal threats to a functional societal structure: male competition for females. Male competition can pose a significant threat to societal stability, as the near-equal ratio of males and females being born necessitates an ideal distribution of mates to maintain peace. This is challenging in practice because females have their own sexual preferences and may prefer to share a high-quality mate with other females rather than settle for a less desirable one, such as one prone to abusive or neglectful behavior. Consequently, a proportion of males will not be able to secure a partner, potentially leading to frustration and social unrest.

Religions and cultural norms have historically played a significant role in maintaining peace by regulating the distribution of women among men. Many religious doctrines advocate for monogamy as a means to control and organize sexual relationships within a community. These norms help to mitigate male competition and promote social stability. Although some religions permit polygyny, it is typically practiced only by a small minority of wealthy males who can afford to sustain multiple wives.

In modern societies, polygyny is typically outlawed in favor of monogamy. This helps maintain social stability by encouraging a balanced distribution of women among men, reducing male competition and potential conflict over mates. However, unlike religious societies, modern societies do not coerce women into forming pairs with men. Many women will choose to remain single rather than settle for a substandard mate. Consequently, some men may find themselves unable to secure a partner in a free society, which can negatively impact social order and their mental well-being. This imbalance can lead to frustration and social unrest, as well as feelings of isolation and dissatisfaction among those men who remain unpaired.

Given that men have historically devised various doctrines or moral systems to reduce male competition over females, it is conceivable that this is one of the greatest threats to an emerging society. It would not be surprising if nature attempted to address this challenge through natural selection and the biological advancement of our species. One possible candidate, that potentially reduces male competitiveness but clearly does not enhance the individual's reproductive fitness, is same-sex attraction. This trait is observed globally and across all cultures. Unlike ADHD, same-sex attraction does not seem to be inherited directly but tends to occur in a perfectly random manner. It appears that all humans have an equal propensity for being born with a preference for same-sex relationships, regardless of their family line or genetic makeup. This universality suggests that same-sex attraction is a desired random variation within the human population.

The evolutionary path for same-sex attraction is not well established, but logically, it must confer a survival advantage to the group as a whole rather than to the individual, since mating with the same gender does not produce offspring, making it a dead end for the lineage. By examining the preference for same-sex relationships through the lens of female distribution over males, we may find clues to the evolutionary fitness of this trait.

Historical records indicate a consistently greater acceptance of male-male sexual relationships compared to those between females. One can reasonably assume that such attitudes were also prevalent in prehistoric times. Furthermore, the physical dominance of males in human populations necessitates that females strategically bond with males for their survival, irrespective of their sexual preferences. The desire to have offspring also does not change with a female's sexual preferences, making strategic bonding appealing for those who want to procreate.

Since females are more inclined to bond strategically despite their sexual preferences, sexual fluidity in humans will create an excess of available females and a lack of men who are hormonally attracted to them. By reducing the number of males competing for females, male same-sex relationships can help mitigate societal tension. This dynamic can lead to a more harmonious social structure where conflicts over mating opportunities are less frequent. Thus, male same-sex relationships may contribute to the overall stability and cohesion of the group, providing an indirect evolutionary benefit despite the reproductive disadvantage for the individual having that preference.

Paradoxically, while male same-sex relationships might be nature's way of fulfilling a purpose traditionally managed by religion, many modern religious doctrines ban the practice.

The objective of this chapter is to underscore the differences in how ADHD and same-sex attraction are inherited. ADHD is usually inherited directly from parent to child, rather than being randomly distributed. If the survival advantages of ADHD were more significant at the population level rather than the individual level, we wouldn't expect it to follow this lineage-based inheritance.

If a population gained significant short-term advantages by having a certain percentage of individuals with ADHD, nature might ensure that the ideal percentage was maintained by having ADHD traits appear randomly in the population, as is the case with same-sex attraction. Achieving an appropriate balance of individuals with lineage-based inheritance requires a harsh, precise, and never-ending selection process to avoid over- or under-expression of the trait. In contrast, a random distribution would ensure that the feature appears at a beneficial level within the population, regardless of parental genetics. This approach would allow the trait to confer its advantages to the group while maintaining genetic diversity without stringent evolutionary control.

Understanding the conclusion of this chapter requires a fair amount of mental gymnastics, and it's not strictly necessary to fully grasp it before proceeding to the next chapter. The overall point is that ADHD is likely to confer some form of individual advantage, given that it is inherited from parent to offspring. In previous chapters, we've discussed how ADHD features such as novelty-seeking may also have population-level benefits, but it must be understood that a feature can be beneficial for humanity without necessarily driving evolution and the biological advancement of our species.

For example, assume that ADHD traits were instrumental in the human invention of the alphabet. That would have been a significant event for humanity, but it does not specifically promote ADHD traits since the invention quickly spread across the population and could benefit anyone, including those without ADHD. Similarly, assuming that ADHD traits were instrumental in the European discovery of America, it does not specifically promote ADHD traits since people without ADHD could also move there or otherwise draw advantages from the opportunities it created. Consequently, these advancements do not place individuals with

ADHD ahead of the general population. For a trait to have evolutionary significance, it can't benefit everybody; it must benefit only a few, or the one, so that some features are promoted over others.

In the case of same-sex attraction, the trait would continuously and repeatedly enhance the stability of small groups of humans that included these individuals, giving them a solid survival advantage over neighboring villages. Over a very long time, this process may change our biology so that it becomes natural for a proportion of males to form pairs with other men instead of competing for the limited selection of females.

However, many of the most significant human advancements were not recurring events but occurred only at singular points in history, such as the invention of the alphabet. These events are not direct participants in evolution because they are too rare, and the benefits are not specific enough to promote a certain biological feature in the population. In contrast, early reproduction offers a strong and immediate benefit that is consistently repeated across numerous individuals in the population. This ongoing advantage promotes the spread of one's genes over those of peers, making early reproduction a powerful evolutionary force that could dominate the evolutionary path of the species.

This does not imply that ADHD traits are universally disadvantageous to survival. There are many examples where novelty-seeking was advantageous to the short-term survival of a group, but likely just as many instances where it led to their downfall, potentially negating any survival benefits conferred by this trait. Since early reproduction appears to be a more consistent advantage for many individuals with

ADHD, it is conceivable that nature's intention is to promote early reproduction rather than conferring survival advantages at either the individual or group level.

The Female Life Cycle

Many women with ADHD are highly functional in their youth, resulting in a lack of diagnosis. However, as they age, the increasing demands of life and hormonal changes can exacerbate ADHD symptoms. This gradual worsening can lead to significant challenges, ultimately becoming a severe impairment that may lead to an adult ADHD diagnosis. Menopause, in particular, can introduce new challenges and make existing symptoms worse.

While the link between ADHD and hormones is well-known, the reasons for the significant changes in ADHD symptoms throughout a woman's life remain unclear. It is uncertain whether these changes occur by chance or nature's design. The theory that ADHD shortens the generational interval provides an experimental framework to interpret these patterns.

Assuming that nature's supposed purpose for ADHD is to drive early motherhood, being functional at a young age is crucial for the survival of offspring. If young women with ADHD exhibited the same presentation and degree of impairment as their male peers, their children could be at risk of neglect and potential death. Therefore, it is reasonable to suggest that ADHD might manifest differently in young women for the purpose of enabling them to be functional caregivers from an early age.

As a woman approaches menopause, her impairment from ADHD is no longer of interest to evolution. As grim as it may seem, there is no selective pressure favoring traits in post-menopausal women. Since a post-menopausal woman can no longer procreate, having favorable traits does not enhance her ability to pass on those genes, which is necessary for evolution to favor them.

Given that a female's ADHD symptoms worsen as she ages and the impact of natural selection decreases, it is possible that her true level of impairment is greater than it seems, similar to the impairment seen in males with ADHD. If so, the better functionality observed during the reproductive years might suggest that young females have natural protective mechanisms that counteract ADHD impairment to protect the offspring from being neglected.

Imagine that evolution could only tolerate the development of ADHD traits if these traits were minimized in women during their fertile years to protect their offspring from maternal dysfunction. Since the female body naturally produces fertility signals, primarily through estrogen, natural selection could exploit these signals to limit ADHD expression. Although hormonal mechanisms are usually more complex, the concept of estrogen regulating ADHD symptoms serves as a placeholder while we explore this theory. In this context, estrogen would not have a genuine purpose beyond acting as a guide signal for evolution to identify fertile women, ensuring that ADHD traits are down-regulated in a gender-specific manner to safeguard the offspring.

The idea that ADHD impairment is influenced by low estrogen levels aligns with the pattern that males are more severely impaired and that this impairment is worst during childhood and adolescence when male estrogen levels are very low. Male estrogen levels increase with age and stabilize around the age of 20. However, since dopamine levels also increase with age, the improvement in ADHD symptoms in men might be more related to dopamine levels. Nonetheless, the pattern of estrogen levels in men aligns with the observed patterns of male ADHD impairment.

Moreover, the suggestion that nature employs estrogen as a guide for ADHD aligns with the observed variations in ADHD symptoms among women during the menstrual cycle. This correlation is extensively documented, drawing from the experiences of millions of women with ADHD. While this isn't novel, this book uniquely explores the evolutionary reasons behind these patterns. The monthly pattern of ADHD symptoms would simply be a side effect of nature's intention to identify and promote fertile women to safeguard their offspring from significant dysfunction.

Suppose that the intensified ADHD symptoms during the menstrual cycle are caused by the proposed evolutionary mechanism: nature deliberately induces executive function disorder in a woman to deter her from pursuing personal dreams and goals, encouraging a simpler and less organized lifestyle favorable to early reproduction. To shield her offspring from maternal dysfunction, nature regulates the disorder using female fertility hormones as guiding signals. If this holds true, one cannot dismiss the possibility that premenstrual syndrome (PMS) might be driven by the same evolutionary mechanisms. Following this reasoning, nature may have intentionally induced PMS mood disorder in a woman to divert her from pursuing—or succeeding with—her personal dreams and goals, promoting a simpler and less organized lifestyle favorable to early reproduction. PMS could potentially be down-regulated by female fertility hormones to protect the offspring from continuous maternal dysfunction and to enhance the mood during the fertile phase of the cycle.

As far-fetched as it may seem, there is sound logic to this idea. The notion that nature impairs humans to promote faster reproduction could potentially explain the emergence of mood disorders such as PMS and bipolar disorder. Similar to ADHD, mood disorders might gain evolutionary fitness by possibly driving early parenthood through impairment. This reproductive advantage could have helped these disorders emerge and persist despite imposing significant mental challenges on those who suffer from them.

It is essential to acknowledge the uncertainty of this line of thought. While it suggests a possible evolutionary mechanism behind mood disorders, it is crucial to approach it critically. These disorders are highly complex, making it difficult to predict their evolutionary path. Readers should consider that the reasoning in this chapter may be flawed and should not take it as absolute truth. Nevertheless, it highlights a grim reality often overlooked in evolutionary science: nature doesn't always have our best interests at heart. Since humans have achieved near-perfect survival rates through adaptability and ingenuity, we are no longer subject to the stringent selective pressures that would protect our mental fitness in nature. On the contrary, traits that accelerate and increase the rate of reproduction may be promoted, even if they contribute to our misery.

Genetics and ADHD

If the hypothesis of a shortened generational interval holds true, it suggests that any gene producing ADHD-like features could independently gain an evolutionary advantage, with some of these features being impairing enough to qualify as a disorder. Over the course of thousands of years, this could lead to a spectrum disorder, where various ADHD-like genetic traits have evolved separately, adding complexity to the condition.

The concept of ADHD as a spectrum disorder is consistent with observed patterns of the condition. ADHD manifests in diverse forms and degrees of impairment, with symptoms and severity varying widely among individuals. Research suggests that ADHD has a complex genetic foundation, involving multiple genes that influence various aspects of the disorder, including dopamine regulation and brain development. Despite the diverse ways ADHD manifests, research indicates a substantial genetic influence, with estimates placing its heritability between 70% and 80%, rather than being predominantly random or environmentally driven.

The groundbreaking study by Ditte Demontis, Raymond K. Walters, and their team, published in *Nature Genetics*, sheds light on the genetic factors influencing ADHD. They analyzed the genetic data of over 20,000 individuals with ADHD and more than 35,000 without the disorder to identify specific genetic regions associated with an increased risk of ADHD. (Nature Genetics[5]).

The study measured the genetic correlation between ADHD and 258 different traits. Some of these correlations are expected due to known comorbidities with ADHD, such as body weight and depression, while others are less obvious. The top ten significant correlations identified are as follows:

1. **Age of Mother at First Birth**: Approximately 65% genetic correlation, indicating that genetic factors linked to ADHD are strongly associated with becoming a young parent.

2. **Educational Attainment**: About -52% genetic correlation, meaning that genetic factors increasing the risk of ADHD are also associated with lower educational achievement.

3. **Smoking History**: Around 50% genetic correlation, suggesting that genetic factors contributing to ADHD also increase the likelihood of having been a smoker.

4. **Years of Education**: Roughly -50% genetic correlation, highlighting a strong link between ADHD and fewer years of formal education.

---

5 Discovery of the first genome-wide significant risk loci for ADHD. Retrieved from https://www.biorxiv.org/content/10.1101/145581v1

5. **BMI (Body Mass Index)**: Approximately 40% genetic correlation, indicating that genetic factors for ADHD are also linked to higher BMI.
6. **Depressive Symptoms**: About 35% genetic correlation, showing a significant overlap between genetic factors for ADHD and depression.
7. **Risk-Taking Behavior**: Around 33% genetic correlation, suggesting a strong genetic link between ADHD and a propensity for risk-taking behaviors.
8. **Obesity**: Roughly 30% genetic correlation, indicating a notable overlap between genetic predispositions for ADHD and obesity.
9. **Cognitive Performance**: About -30% genetic correlation, showing that genetic factors for ADHD are associated with lower cognitive performance.
10. **Age at First Sexual Intercourse**: Approximately 28% genetic correlation, suggesting that genetic factors linked to ADHD also influence the age of first sexual activity.

In light of the book's hypothesis, the results are particularly striking. The single most correlated factor with ADHD is that of becoming a young parent. This link is rarely given consideration as an independent factor, possibly due to bias. It may be assumed that ADHD is primarily linked to social poverty and low education, which are traditionally seen as precursors to early parenthood. However, the strong genetic correlation suggests that ADHD itself might be a significant factor in the timing of parenthood, independent of these socio-economic variables.

The 65% correlation to young parenthood is indeed significant, outstripping other measured features in the study. However, it may still understate the true impact due to a curious paradox.

Figure 3 shows the relationship between parental age and the risk of ADHD in offspring, as observed in the meta-analysis conducted by Min et al. The study found a non-linear relationship between parental age and the risk of ADHD in offspring. The lowest risk of ADHD in offspring was observed when parents were aged 31-35 years old, while both younger and older parental ages were associated with higher risks (IJERPH[6]).

---

6 Parental Age and the Risk of ADHD in Offspring: A Systematic Review and Meta-Analysis. Retrieved from https://www.ncbi.nlm.nih.gov/pmc/articles/PMC8124990/

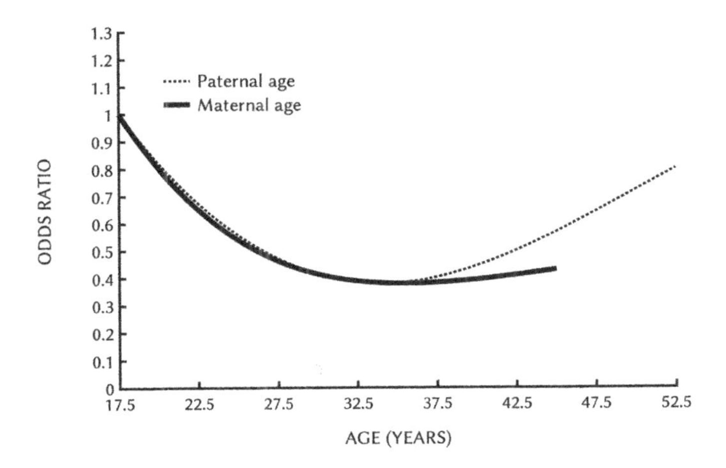

Figure 3: Parental age vs. offspring ADHD risk

This may not be as surprising as it initially seems. Many consider the mid-twenties to early thirties the ideal age for parenthood, assuming life goes as planned. However, for individuals with ADHD, life often deviates from the plan. Starting a family within a desired timeframe necessitates a robust set of executive function skills, such as planning your family's trajectory, following through on those plans, and maintaining stable long-term partnerships. These aspects can all be very challenging for a person with ADHD, making parenthood more likely to occur randomly throughout their lifetime. Consequently, the increased ADHD risk associated with very young mothers and older mothers may possibly be explained by their genetic disposition to have children outside of the optimal timeframe.

Similar U-shaped patterns for parental age can be observed for various disorders, including autism spectrum disorder

(ASD), schizophrenia, and bipolar disorder, suggesting that children of very young and older mothers may have a higher risk for mental disorders. Traditionally, this is attributed to factors related to the mother's age. For younger mothers, factors such as lower socioeconomic status and biological immaturity are said to contribute to higher risks. In older mothers, increased risks are attributed to higher rates of chromosomal problems, genetic mutations, and complications during pregnancy. Nevertheless, there is no solid proof that a mother's age causes mental disorders in a child, making these explanations largely speculative. Additionally, these theories might be influenced by societal biases against mothers outside conventional age ranges.

Given the insights shared in this book, a simpler explanation emerges: It takes significant mental fitness to start a family within the optimal time frame. If ADHD can distort the generational interval, similar patterns might be observed for other mental disorders. It is conceivable that parents who manage to start a family between the ages of 25 to 35 are more likely to be mentally fit, a feature which may be passed on to their children. Consequently, these children's superior mental health may reflect their parents' genetic fitness rather than their age at childbirth. Conventional individuals tend to have their children within a narrowly defined and carefully planned time frame to align with societal expectations. If inherited personality traits, such as those associated with mental disorders, prevent individuals from adhering to this conventional timeline, their offspring's birth distribution will be more random. As a result, when the risk associated with these genetic dispositions is measured against the general population, it produces the illusion of an elevated risk for children born to both very young and older parents.

In reality, the U-shaped pattern may not necessarily signify an elevated risk but rather a family-planning anomaly: parents with mental disorders may have their offspring more randomly because they can't organize their childbirths as effectively around the 30-year mark. The U-shaped pattern of their disorder may lack a biological connection but could simply reflect the bell-shaped age distribution of planned parenthood among conventional parents. Therefore, the biological concerns related to having offspring too early or too late in life may be overstated.

As with many assertions in this book, this idea remains a theory. However, it offers a logically coherent explanation that might be more appealing than blaming the mother's age for mental disorders. It seems counterintuitive that newborns would be healthier when born to mothers aged 31-35, given that this was the expected lifespan of a Stone Age female rather than the prime reproductive age.

Epilogue

The name "ADHD" highlights three primary aspects of the disorder: hyperactivity, impulsiveness, and inattentiveness. Although traditionally viewed as distinct features, they share a common underlying theme—they all represent different forms of fast-forwarding through life, rather than experiencing it at its natural pace. A hyperactive person is constantly restless, always looking for the next activity, seemingly living life at a faster pace than their surroundings. An impulsive person refuses to delay gratification, taking the quickest and most direct route toward a set goal. An inattentive person skips essential parts of life, missing out on significant opportunities and experiences along the way. This tendency to skim through life might be nature's strategy to minimize generational intervals, ensuring that individuals with these traits avoid self-serving activities that delay reproduction.

Russell A. Barkley, a prominent psychologist and researcher in the field of ADHD, describes ADHD as a disease of "time blindness." He emphasizes that individuals with ADHD have significant difficulty perceiving and managing time, which fundamentally affects their ability to plan, organize, and regulate their behavior in accordance with future goals.

While this description is insightful, we might need to consider the possibility that ADHD may not be primarily a matter of time blindness, but rather a "family-planning disorder," where time blindness is simply one of many core features for nature to fulfill that goal.

By looking at ADHD through this lens, the variety of ADHD presentations fits under the same umbrella, presenting a framework that might explain the existence and propagation of these human features. Nature equipped us with impulses for the purpose of acting on them, and it does not help if we have an unlimited ability to withstand them all, prioritizing our life experiences over the creation of new life. It's possible that, as humans became too strategically minded and self-serving for our own good, nature may have stepped in to ensure we would not stray from our mission to procreate. By throwing a wrench in the gears of the human mind, natural selection may have invented lineages that fast-track the human life cycle, gaining an unexpected evolutionary advantage.

ADHD may not be an impairment in the sense that it's nature's mistake. Rather, it might be a manifestation of nature fighting back against the human mind. Humans have been the world's top predator for a million years, achieving almost perfect survival rates by using our superior intelligence to adapt to every challenge that crossed our path. Since surpassing nature one million years ago, our trajectory has long been dominated by sexual selection rather than survival selection. Instead of leaving it to nature to shape our physique by preserving the fittest and strongest individuals, our men have bred their women while our women have bred their men, judiciously selecting the mate that attracts us the most in the spur of the moment.

In that process, we have developed brains so oversized that our babies need several months of neck-muscle training to hold their heads up. We have lost our fur, our built-in protective insulation against the weather, in favor of naked bodies that dress fancy and wear ornaments to attract mates. Over a long-term process of men selecting women who develop breasts outside of childbearing, our females have acquired inconvenient fat deposits that mimic lactation, impairing their ability to move freely. Meanwhile, women have selected the tallest men with the largest sexual organs, resulting in men who are almost two meters tall and with penis sizes ten times that of a gorilla. In essence, humans are peacock monkeys, spawned through a million years of undisturbed sexual selection where we have sidestepped nature in favor of our choices of useless features that meet the demands of our mating rituals. In that process, we risk drowning in our own reflection, losing track of what the mating process was actually about.

Evolution may have opted to win the battle against the human mind by creating a lineage of nature's warriors who are more attentive to our natural impulses, ensuring that some of us preserve the primary mission of procreation. Nature designed us to survive and procreate, not for family planning. The ability to plan against our own procreation was always an unwanted skill by evolution, as the natural order is to live in the moment. The strategic foresight of an animal was never supposed to span more than a few weeks or months to ensure our immediate survival. It violates the procreation contract to create a master plan that forecasts an entire lifetime and delays childbirth until the individual decides the time is right.

In conclusion, ADHD traits may serve as a reminder of our primal instincts, bringing us back to the fundamental purpose of life: to survive and reproduce. By understanding these traits within the broader context of human evolution, we can appreciate the complex interplay between human willpower and human nature. Embracing this perspective allows us to see ADHD not merely as a disorder, but as nature's representative in our battle against ourselves. It is possible that, in an attempt to protect us from the consequences of conquering our own reproduction, nature made us a diverse species of highly intelligent but relentlessly novelty-seeking animals that ended up conquering the world instead of conquering our own minds.

Made in the USA
Las Vegas, NV
05 August 2024